BLACK HOLES!

The little black hole at the edge of the galaxy.

WRITTEN & ILLUSTRATED BY: NICOLE CUNKLE

Made for and dedicated to Sebastian, a curious kid with many questions.

www.PetRockPress.com

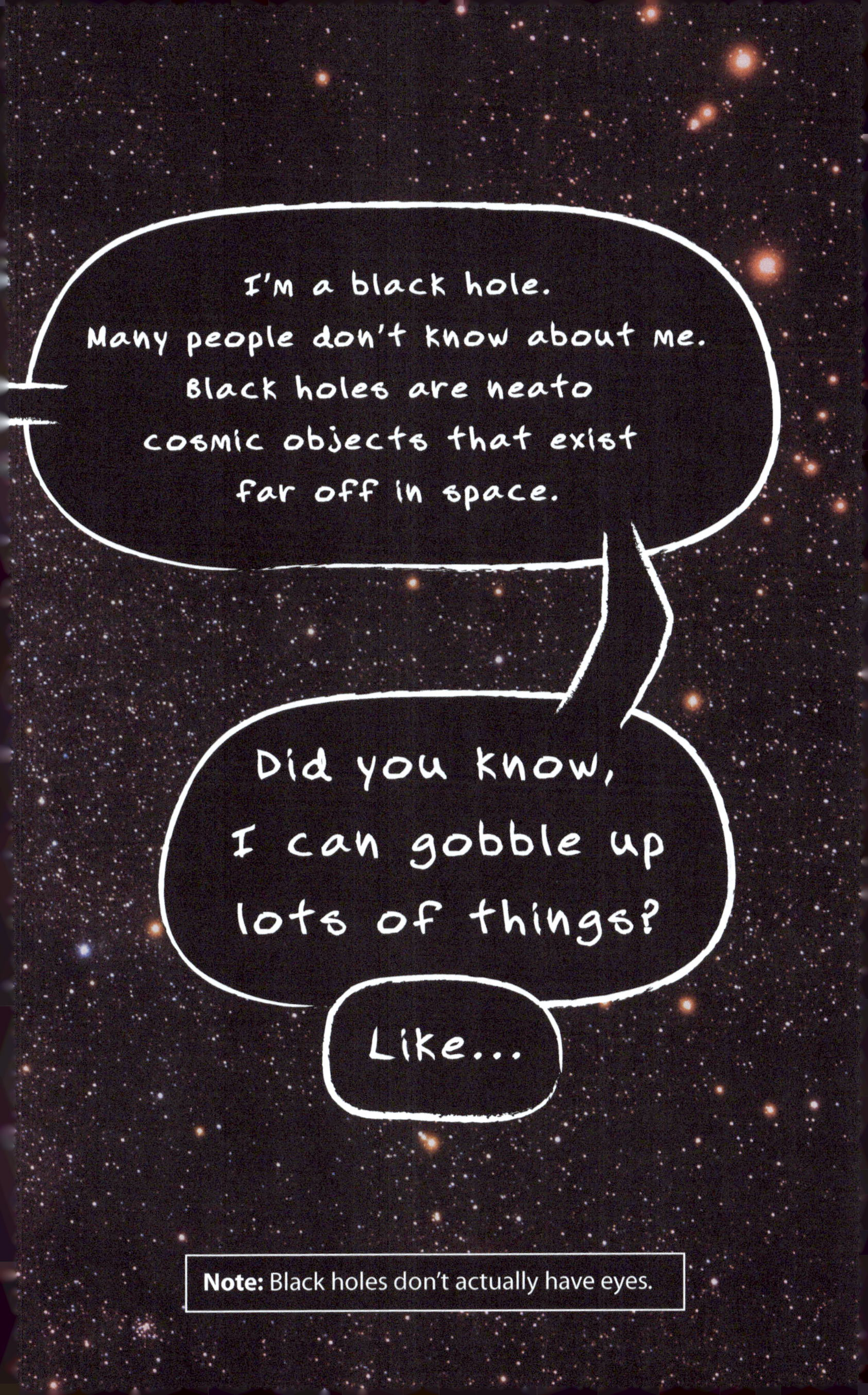

Note: Black holes don't actually have eyes.

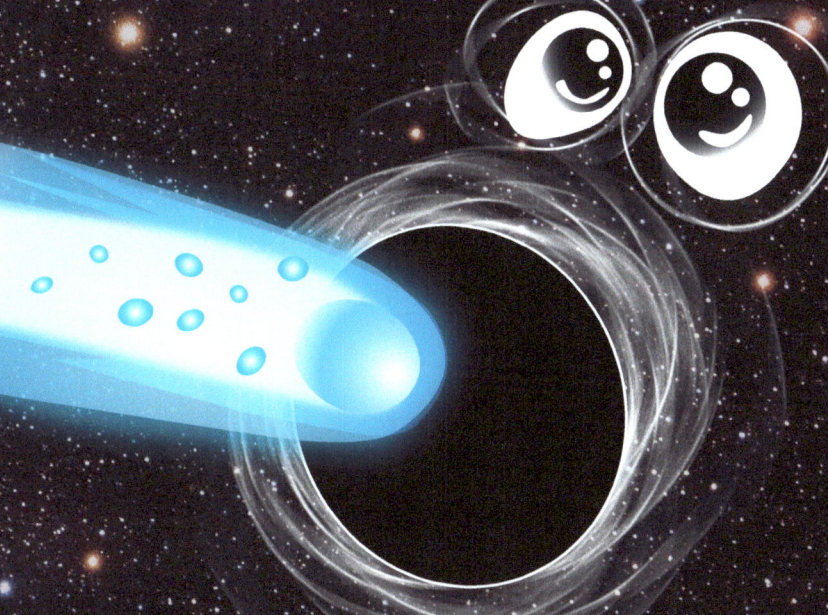

EVEN LIGHT!

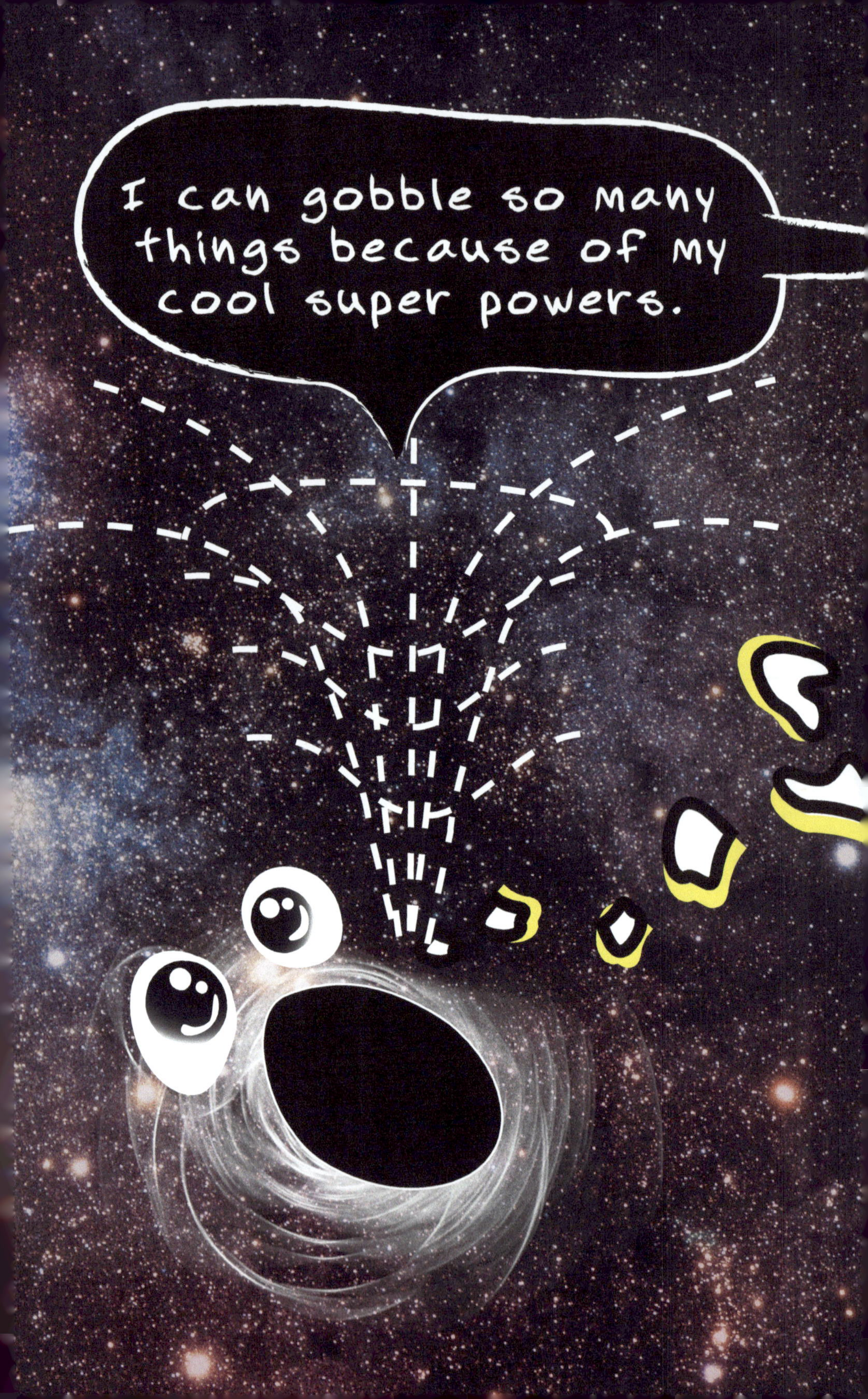
I can gobble so many things because of my cool super powers.

It's the force that makes apples fall to the ground.

Gravity
pulls things past my
Event Horizon!

The event horizon
is like when you go
down a waterfall
and can't
paddle back up.

It's my boundary from which... NOTHING CAN ESCAPE!

After the event horizon, things start to become stretched out...

like a noodle

and this is called...

spaghettification!

With all these cool powers, one day I hope to be a...

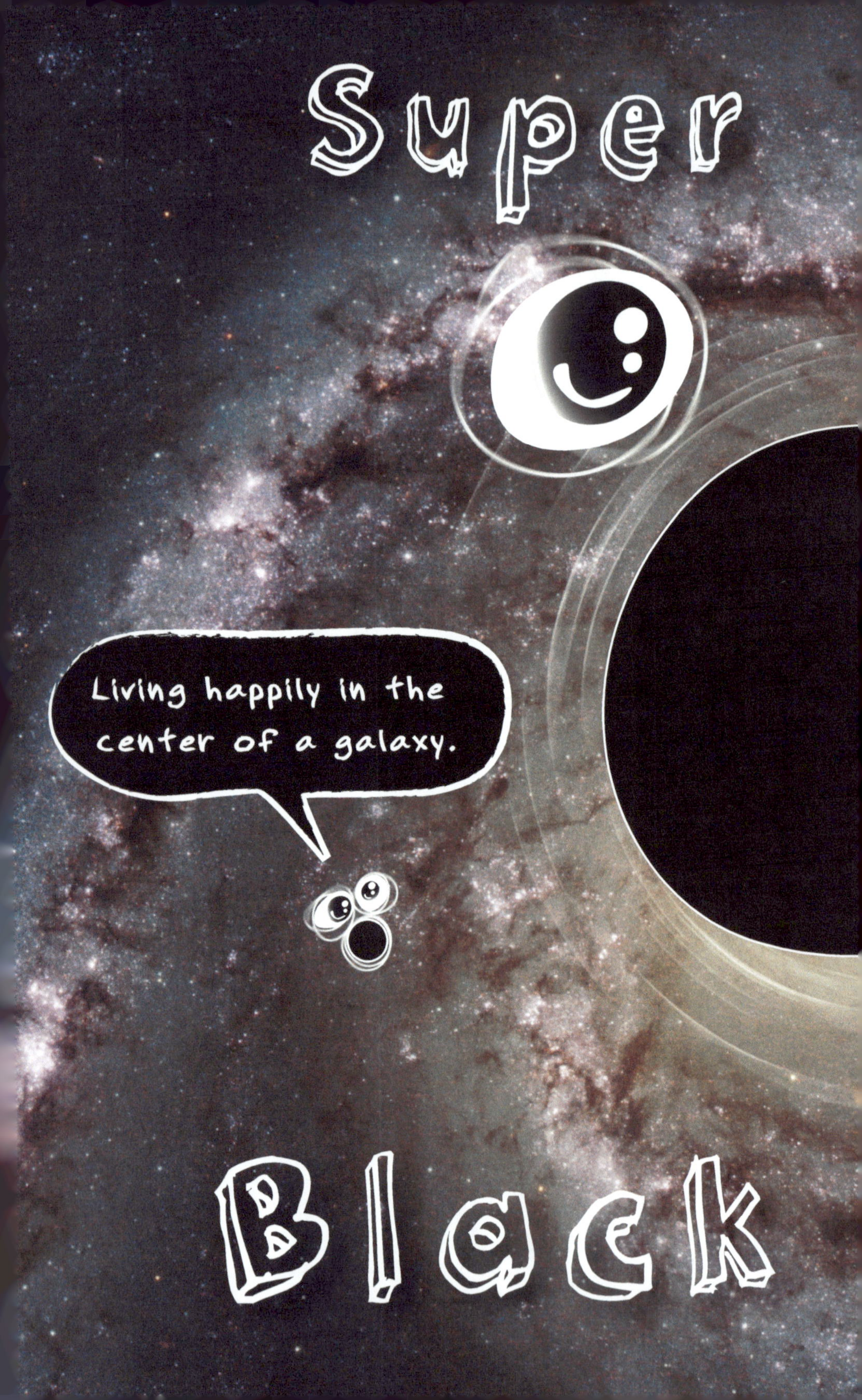
Super
Living happily in the center of a galaxy.
Black

Massive

Hole!

GLOSSARY:

Black Hole
A massive object with a gravitational pull so strong that nothing can escape.

Gravity
The force that makes things fall towards each other.

Density
Is the total amount of stuff smooshed into one space.

Mass
Is a group of parts or elements in no specific shape.

Singularity
A single point in which everything gets smooshed in a black hole.

Infinity
Without having an end.

Event Horizon
A black hole's point of no return. After this, not even light can escape.

Spaghettification
Tidal forces that stretch an object when inside a black hole.

CONTINUED...

Comets
An icy, dusty object in space, that makes a bright tail as it travels through space.

Stars
Atom smashing balls of gas burning far away.

Planets
A large, mostly spherical object in space, which revolves around a star.

Galaxy
A super large group of stars.

Nebulas
Space clouds made of gas and dust - Where stars are born.